DREAM INVENT CREATE

Engineer the World

An introduction to the inspiring possibilities of engineering

SPACE SHIPS, SATELLITES, AND MORE. ★ **WHAT ELSE?** THEY'RE DEVELOPING VEHICLES TO MAKE SPACE **SO COOL** AEROSPACE ENGINEERS DESIGN AND BUILD AIRPLANES, JET FIGHTERS, ROCKETS, SHIPS, OR PLAN A FUTURE HUMAN COLONY ON THE MOON.

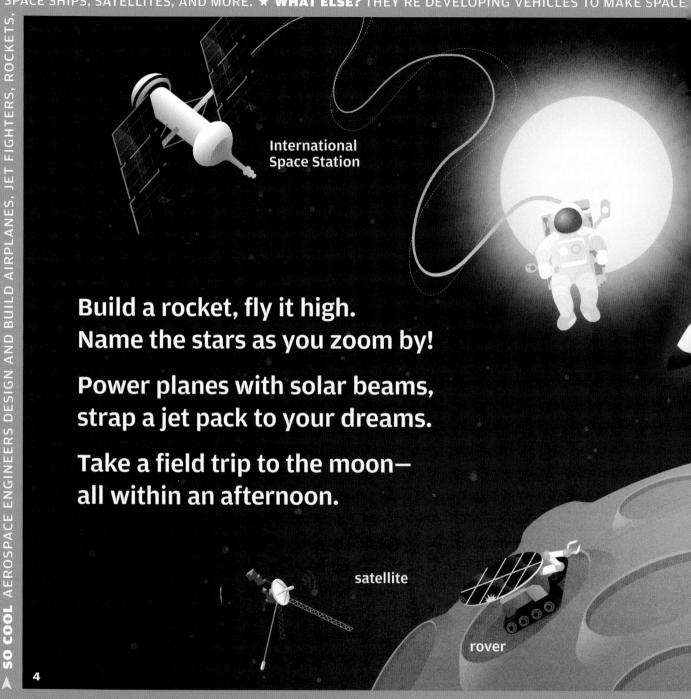

International Space Station

Build a rocket, fly it high.
Name the stars as you zoom by!

Power planes with solar beams,
strap a jet pack to your dreams.

Take a field trip to the moon—
all within an afternoon.

satellite

rover

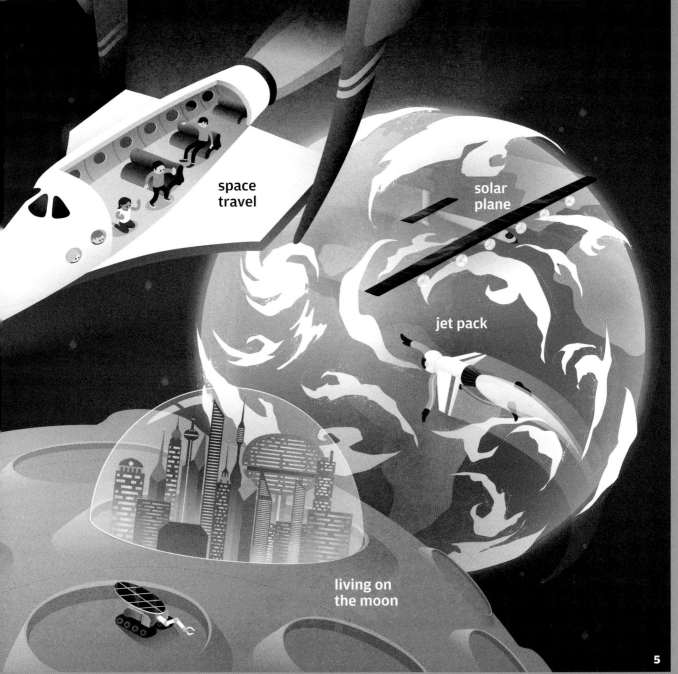

HIGH-QUALITY, NUTRITIOUS FOOD. ★ **WHAT ELSE?** THEY'RE DESIGNING VERTICAL FARMS TO BE BUILT IN CI

SO COOL AGRICULTURAL ENGINEERS DEVELOP NEW WAYS TO GROW, HARVEST, AND DISTRIBUTE

Grow food enough for every belly from New Hampshire to New Delhi.

Make your veggies more delicious and your burgers more nutritious.

Plant farms in unexpected places for cleaner, greener urban spaces.

greenhouse on moon

irrigation

farming

NEW TYPES OF PLANTS THAT CAN BE MADE INTO BIOFUELS FOR CARS TO USE INSTEAD OF GASOLINE.

CAN USE TO DIAGNOSE AND TREAT PATIENTS. ★ **WHAT ELSE?** THEY DESIGN PROSTHETICS SO ATHLE

Help physicians treat their patients with new and better medications.

Build a laser more precise than the most skilled surgeon's knife.

Design advanced robotic limbs so everyone can run and swim.

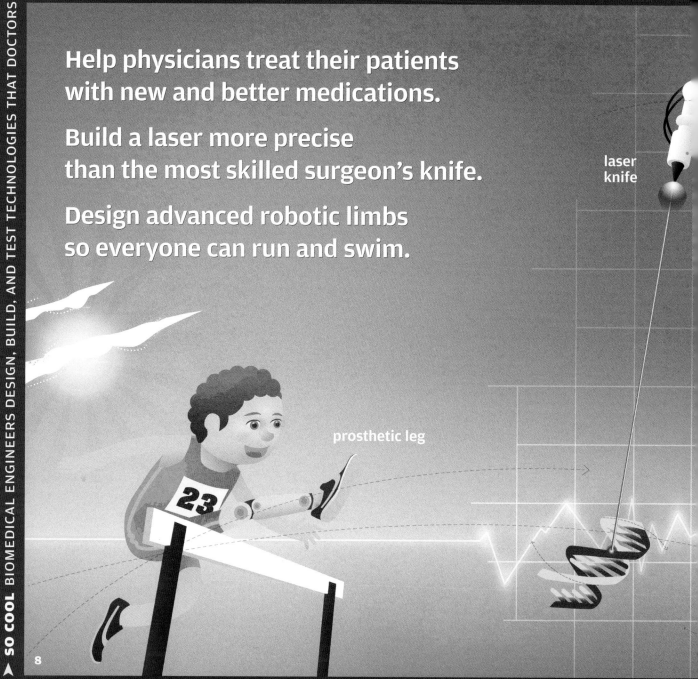

laser knife

prosthetic leg

SO COOL BIOMEDICAL ENGINEERS DESIGN, BUILD, AND TEST TECHNOLOGIES THAT DOCTORS

ARTIFICIAL ORGANS LIKE HEARTS, KIDNEYS, AND LIVERS, AND GROWING TISSUES LIKE SKIN AND BONE.

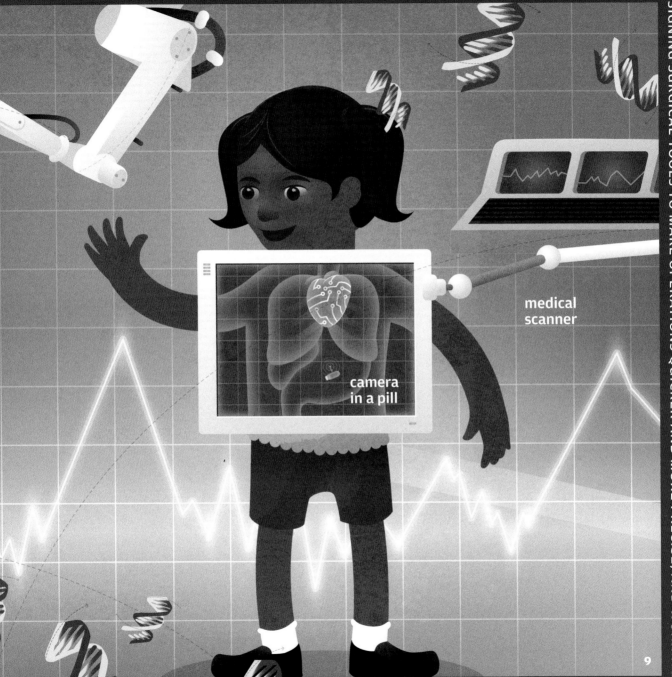

Concoct delicious flavored drinks or brew some color-changing inks.

Build everlasting batteries to power phones and cars with ease.

Make paint that gobbles air pollution — What a clean and green solution!

...STICS FOR SMARTPHONES, COLORED DYES FOR CLOTHING, AND LONG-LASTING PAINT FOR BUILDINGS. ★ **SAVING THE PLANET** CHEMICAL ENGINEERS WORK TO MAKE CHEMICAL PROCESSES USE LESS ENERGY AND GENERATE LESS WASTE. ★ **TELL ME MORE** THEY MIGHT HELP TO DESIGN AND MANAGE...

medicine

batteries

paint

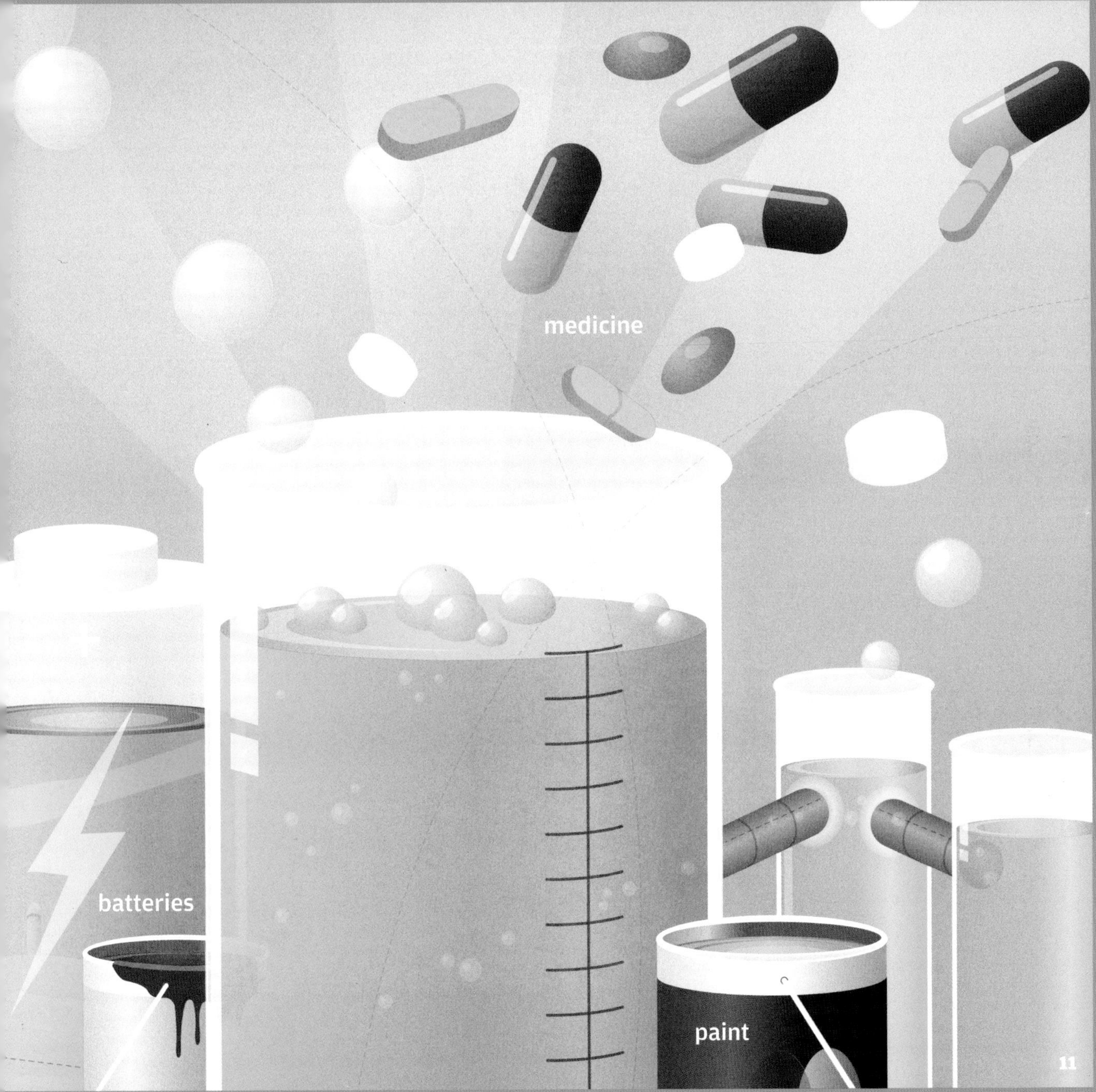

Design a city grand and clean
with high-speed trains and spaces green.

With graceful bridges stretching wide
and sturdy dams to stem the tide —

A place where people work and play.
What cities can you dream today?

SOCIETY, LIKE CONSTRUCTING HYDROELECTRIC DAMS THAT PRODUCE ELECTRICITY AND SUBWAY SYSTEMS THAT CARRY PEOPLE ACROSS TOWN. ★ **SAVING THE PLANET** CIVIL ENGINEERS MIGHT FIGURE OUT WAYS TO HEAT AND COOL A SKYSCRAPER USING LESS ENERGY. ★ **TELL ME MORE**

SMARTPHONES, AND OTHER ELECTRONIC GADGETS WE USE TODAY. ★ **WHAT ELSE?** THEY OFTEN DEVELOP SOFTWARE TO CONTROL ROBOTS, AND WORK ON CYBERSECURITY TO PROTECT INFORMATION.

SO COOL COMPUTER ENGINEERS DESIGN THE SOFTWARE AND HARDWARE FOR COMPUTERS,

Program all your favorite toys.
Create fun games for girls and boys!

Build the next amazing app
to download with a finger tap.

Gear up your imagination
for making 3-D animation.

video games

cyber security

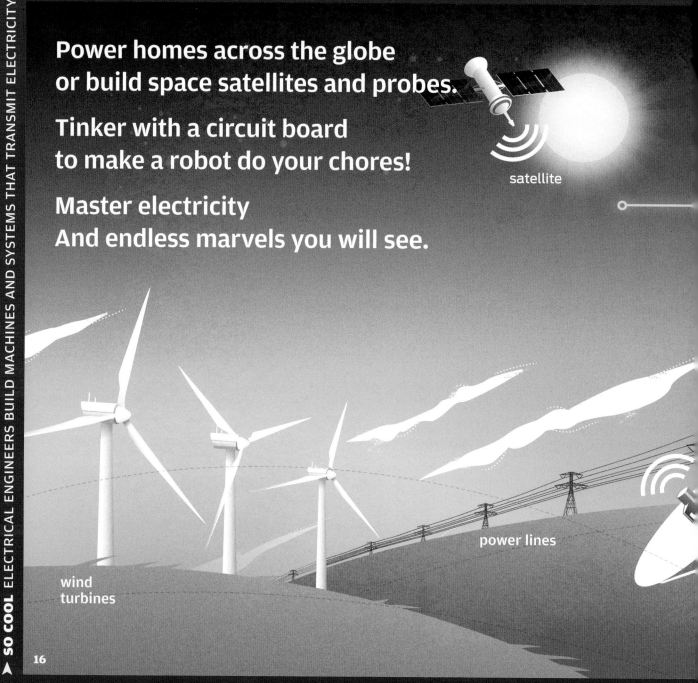

Power homes across the globe
or build space satellites and probes.

Tinker with a circuit board
to make a robot do your chores!

Master electricity
And endless marvels you will see.

satellite

wind turbines

power lines

SO COOL ELECTRICAL ENGINEERS BUILD MACHINES AND SYSTEMS THAT TRANSMIT ELECTRICITY FROM WHERE IT'S PRODUCED TO WHERE IT'S USED. ★ **WHAT ELSE?** SOME ALSO APPLY THEIR KNOW-HOW TO DESIGN SATELLITES AND COMMUNICATION SYSTEMS TO SEND INFORMATION AROUND THE WORLD.

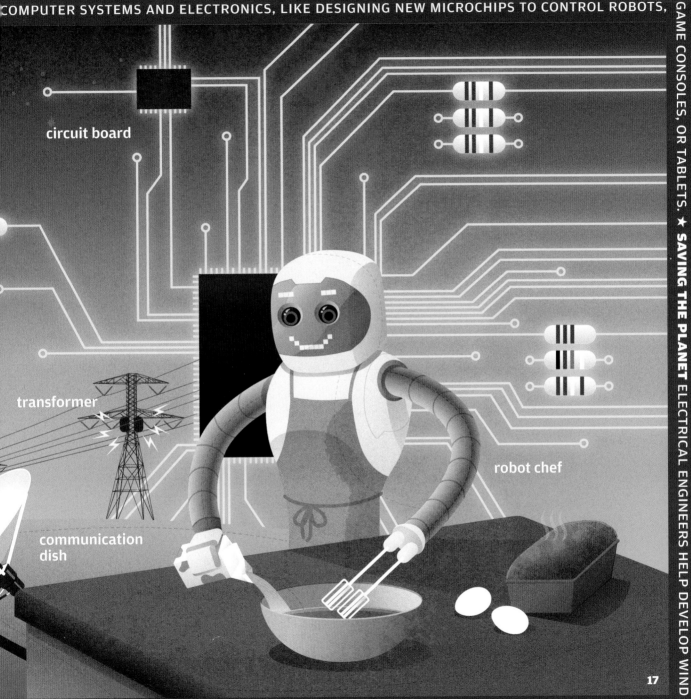

PLANTS, SOIL, AND WATER. ★ **WHAT ELSE?** THEY DESIGN SYSTEMS TO PREVENT AND CONTROL POLLUTI

SO COOL ENVIRONMENTAL ENGINEERS DEVISE SOLUTIONS TO PROBLEMS THAT FACE OUR AIR,

**Keep our planet lush and green
with air that's pure and water clean.**

**Convert old trash to useful things
like purses, clothes, and playground swings.**

**Learn to think sustainably
for the good of bird and bee.**

recycling

UBLIC POLICY TO FIGURE OUT ENVIRONMENTAL SOLUTIONS THAT BENEFIT ALL PARTS OF SOCIETY.

...NSERVE THE EARTH'S RESOURCES, AND SLOW DOWN GLOBAL CLIMATE CHANGE. ★ **SAVING THE PLANET** ENVIRONMENTAL ENGINEERS ARE WORKING ON NEW WAYS TO COLLECT AND SORT WASTE SO THAT MORE OF IT CAN BE RECYCLED. ★ **TELL ME MORE** THEY MIGHT ALSO LEARN ABOUT THE LAW AN...

clean air and land

water treatment

clean water

THAT PEOPLE USE TODAY—INCLUDING TOYS, CARS, AIRPLANES, AND MORE. ★ **WHAT ELSE?** THEY HEL

laser cutter

3-D printer

**Be a business pioneer
with factories to engineer.**

**Robots on assembly lines
that fabricate new toy designs.**

**And make your dreams reality
with 3-D print technology!**

▶ **SO COOL** MANUFACTURING ENGINEERS DESIGN FACTORIES AND SYSTEMS TO MAKE ALL THE STUFF

BLOCKS OF MATTER, SOMETIMES IMITATING ONES FOUND IN NATURE. ★ **WHAT ELSE?** THEY USE CHEMIST

glass that changes color in response to sunlight

sports equipment

See the world at nanoscale and innovation will prevail.

Manipulate small particles to make fantastic articles—

Fabrics light but strong as steel, metals that can break, then heal.

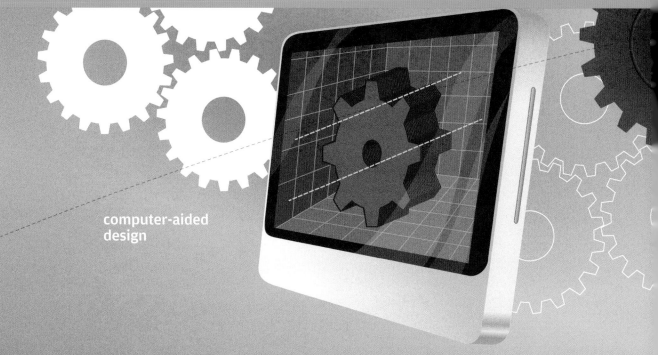

computer-aided design

Build the fastest roller coaster,
a car, a plane—a better toaster!

Software tools can simulate
the awesome things
you will create.

The engine's hum,
the whir of gears
will be like music to your ears.

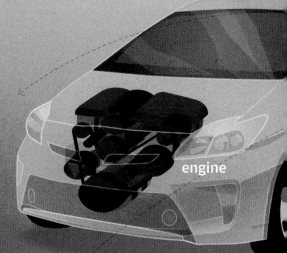

engine

SO PEOPLE WILL HAVE THE RAW MATERIALS TO MAKE THINGS. ★ **WHAT ELSE?** THEY MIGHT DESIGN MINES

mining site

Adventure deep beneath the ground
where precious diamonds may be found.

Bore through mountains and through hills
with dynamite and massive drills.

Use the best newfangled tools
to bring the world its needed fuel.

▶ **SO COOL** MINING ENGINEERS DEVELOP TECHNIQUES FOR GETTING MINERALS OUT OF THE GROUND

THEIR EXPERTISE TO FIGURING OUT HOW TO DIG TUNNELS FOR A NEW SUBWAY SYSTEM UNDERNEATH A CITY.

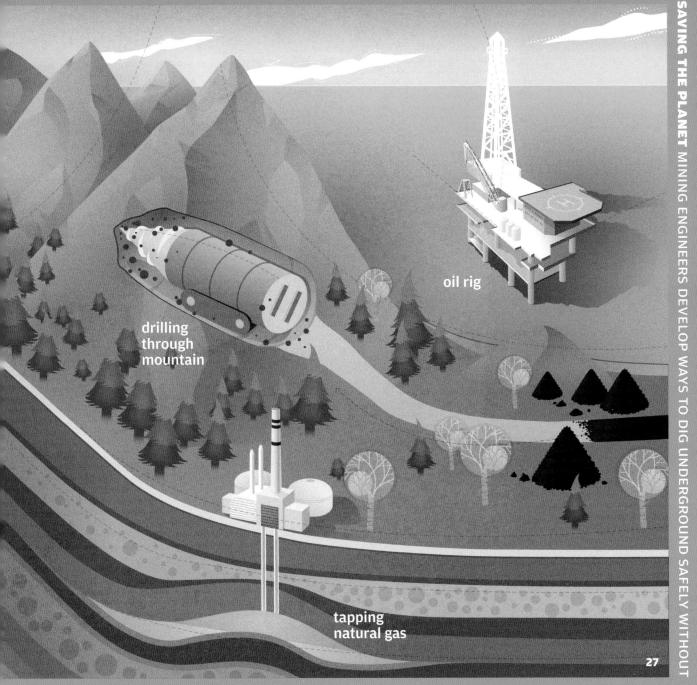

THE WATER AND SUBMARINES THAT DIVE BENEATH THE WAVES. ★ **WHAT ELSE?** THEY ALSO DESIGN MAR

SO COOL OCEAN ENGINEERS DESIGN AND BUILD MARINE VESSELS, LIKE SHIPS THAT SAIL ACROSS

Explore the ocean vast and blue aboard a tanker built by you.

Or dive to depths as yet unseen in a deep-sea submarine.

Guide robots' underwater trips for finding whales and sunken ships.

buoys

underwater turbines

HICLES WHOSE PROPULSION SYSTEMS, CONTROLS, AND INSTRUMENTS WERE ALL BUILT BY ENGINEERS.

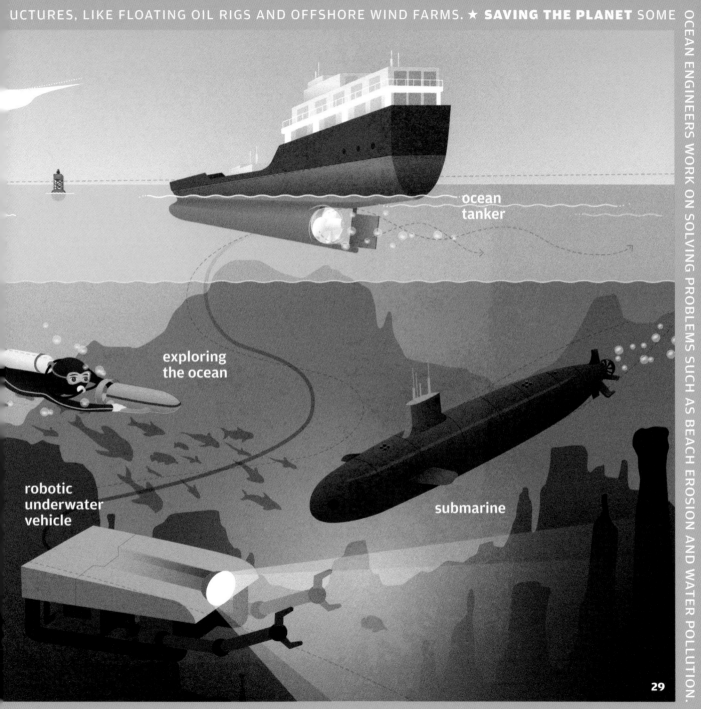

COMPLEX PROJECTS INVOLVING PEOPLE, PROCESSES, GOODS, AND INFORMATION. ★ **WHAT ELSE?** THEY OF

SO COOL SYSTEMS ENGINEERS THINK ABOUT THE BIG PICTURE, FIGURING OUT HOW TO MANAGE

Use mathematics and statistics to calculate complex logistics.

Know who and what goes when and where over land or sea or air.

Manage systems with precision and watch the world dance to your vision!

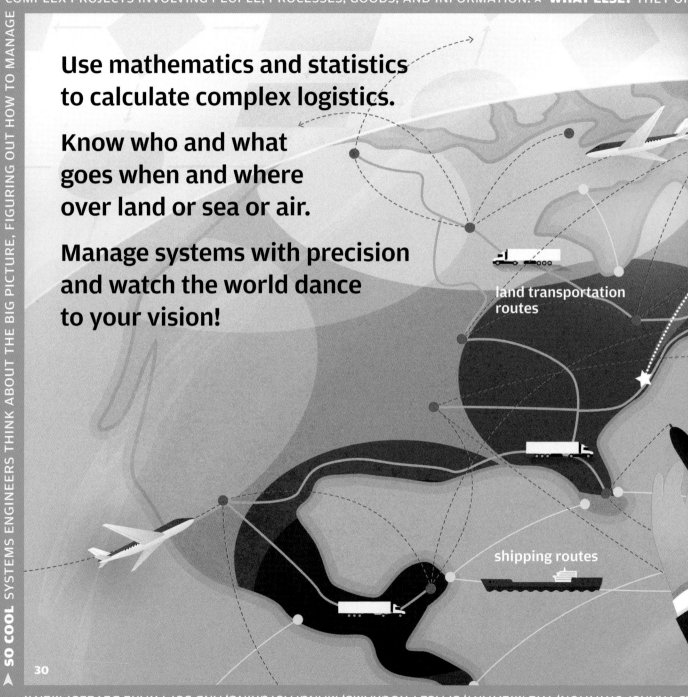

land transportation routes

shipping routes

AS TRANSPORTATION, THE MILITARY, SPACE PROGRAMS, MANUFACTURING, AND SOFTWARE DEVELOPMENT.

DEEP KNOWLEDGE ABOUT LOTS OF DIFFERENT AREAS, AS WELL AS ANALYTICAL AND ORGANIZATIONAL SKILLS. ★ **NEW STUFF** SYSTEMS ENGINEERS MIGHT USE COMPUTER MODELING SOFTWARE TO PREDICT HOW DIFFERENT PARTS OF A SYSTEM WILL WORK TOGETHER. ★ **TELL ME MORE** THEY WORK IN MANY FIELDS,

space launch

air traffic

troop and resource movement

MACHINES TO THE TALLEST SKYSCRAPERS. ★ **WHAT ELSE?** ENGINEERS WORK IN TEAMS WITH PEOPL

So make your big dreams come alive! Help humanity to thrive.

Invent amazing new designs and let your full potential shine.

Create a future bright and new— all the rest is up to you.

▶ **SO COOL** ENGINEERING IS IN EVERYTHING ALL AROUND US, FROM THE TINIEST MICROSCOPIC

★ **TELL ME MORE** THE WORLD ALWAYS NEEDS ENGINEERS. THEY ARE IN DEMAND AND WELL PAID.

...DS LIKE ARCHITECTURE, LAW, BUSINESS, DESIGN, AND MEDICINE TO SOLVE IMPORTANT PROBLEMS. ★ **SAVING THE PLANET** ENGINEERS DO WORK THAT HELPS PEOPLE EVERYWHERE LIVE HAPPIER, HEALTHIER LIVES. ★ **NEW STUFF** THEY NOT ONLY WORK WITH THE LATEST TECHNOLOGIES BUT ALSO INV...

Scavenger Hunt

Can you find these details in the illustrations of this book?

- ☐ a dolphin
- ☐ golf club
- ☐ juice box
- ☐ lunch box
- ☐ a wrench
- ☐ purses
- ☐ greenhouse on moon
- ☐ camera pill
- ☐ medicine
- ☐ kayakers
- ☐ the number 23
- ☐ buoy
- ☐ shipping route
- ☐ oil rig
- ☐ jet pack
- ☐ helicopter landing pad
- ☐ moon rover
- ☐ military tank
- ☐ a pickle
- ☐ vertical farm
- ☐ dump truck
- ☐ tunnel
- ☐ egg beaters
- ☐ 3-D printer
- ☐ ear of corn

Challenge:
- ☐ satellites (2)
- ☐ turbines (2)
- ☐ lasers (2)
- ☐ high-rises (4)
- ☐ robots (6)

Word Search

Can you find the 35 words hidden in here?

```
P M P U T L J T B U F T Q C X S S L I Y
X Y S E S F A C T O R Y O C E A N R C E
M B A T T E R I E S A N E B S S K R S
M T Y R O L L E R C O A S T E R A P P S
C P T A T R A N S P O R T A T I O N I S
A L S N N C H E M I C A L S T L V M S Q
E C A S I D K V S P Y L T P L A S T I C
A M T F F N Y L S U B M A R I N E Z B Z
U H E O I R A J O R O A D S S W S C Z D
Q W L R H I Q T L D D D B R I D G E T W
L C L M R P J V L Q S K Y S C R A P E R
Y P I E G C J Y Q D A F A R M I N G S Q
G N T R C O M P U T E R S Z D I M B I A
W A E J F R M E D I C I N E U D V R M C
M L L A T A R W G R R Z G R E U O R R N
U R Q Z N M B S A D F R E C Y C L I N G
J R A I K K P R T U R B I N E T Y J W D
P A A U D B Y L I C S G B G O Q Y V E S
O R Y O C A R B L C P O O B A F B I M E
T G I V I E T Z F M S A O Z G T I D I E
R D K E V C H S Q Q N R I H S P I E N N
L O I O D B I O F U E L K N N H S O I G
G A R K G V P F R O C K E T T R X G N I
H U S R M P X T L W S R W Q A E J A G N
Q H W E B H R W W R C C J E C L E M L E
P X E D R L U A Z B H A G R B E G E Q B
O I X O Q E N R U R M U V I H Q C A R D
K J N E Y K B E U L U A S I U U D Y P J
```

APPS
BATTERIES
BIOFUEL
BRIDGE
CAR
CHEMICALS
COMPUTERS
ENGINE
FABRICS
FACTORY
FARMING
GEARS
IRRIGATION
LASER
MATERIALS
MEDICINE
MINING
OCEAN
PAINT
PLASTIC
RECYCLING
ROADS
ROBOT
ROCKET
ROLLERCOASTER
ROVER
SATELLITE
SKYSCRAPER
SOFTWARE
SUBMARINE
TRAIN
TRANSFORMER
TRANSPORTATION
TURBINE
VIDEOGAME

Crossword Puzzle

Across
4. Government space agency
5. Replacement leg or arm
6. Type of security for computers and networks.
7. ____ engineers manage complex projects like shipping routes and space launches.
9. Fuel made from plants
10. Building vehicles for underwater exploration would be the work of this type of engineer.
11. A soft drink flavored to taste like blueberry pie might be the work of this type of engineer.
12. A human colony on the moon would be planned by this type of engineer.
13. Engineers are making ____ planes and ships that burn less fuel
14. A skyscraper that can withstand earthquakes and hurricanes would be the challenge of a ____ engineer.

Down
1. Farming in city high-rises
2. Robots are found here, making production faster, cheaper, and safer.
3. These engineers create new types of plastics and metals.
6. These engineers work at places like Facebook and Apple.
8. Field of engineering that works to protect the earth's air, water, and soil.
9. This type of engineer might someday grow bones for transplants.

Answers for all puzzles at www.start-engineering.com

Think About It

1. What did you think an engineer did before you read this book? What do you think now? What surprised you the most?

2. For each page, what other inventions or designs might that type of engineer dream up?

3. How have the inventions on each page changed or improved over time? What were things like before you were born? What might they be like in the future?

4. How would you improve each invention? What would you change or add? Think big! There are no wrong answers.

5. Look around your house or classroom and try to identify items that you think were created by engineers. How have these objects changed or improved over time?

6. Which page of the book is most interesting to you? Why?

7. What would you enjoy about being each type of engineer, and what wouldn't you like?

8. If you were an engineer, what problem do you think you might try to solve?

GET CREATIVE!

1. Make your own rhymes for each page. You could even put them to music or a beat, and record a rap.

2. Research the field of engineering that appeals to you the most and write a short essay with more information on the field.

Now it's YOUR turn to be an engineer!
Take the Marshmallow Challenge

This fun activity* works best in teams of 2-4, racing against the clock to **build the tallest freestanding structure** using 20 sticks of spaghetti, one yard of string, one yard of tape, and one marshmallow.

HERE'S WHAT YOU'LL NEED:

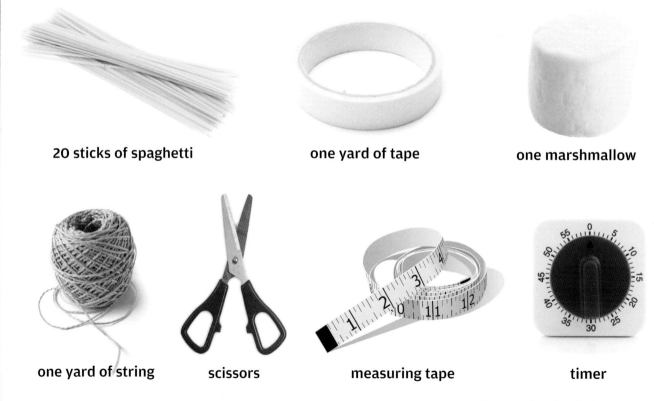

20 sticks of spaghetti

one yard of tape

one marshmallow

one yard of string

scissors

measuring tape

timer

*from marshmallowchallenge.com